Engineering Our World

# How a Satellite Is Built

By Jonathan and Mariel Bard

**Please visit our website, www.garethstevens.com. For a free color catalog of all our high-quality books, call toll free 1-800-542-2595 or fax 1-877-542-2596.**

**Library of Congress Cataloging-in-Publication Data**

Names: Bard, Jonathan, author. | Bard, Mariel, author.
Title: How a satellite is built / Jonathan and Mariel Bard.
Description: New York : Gareth Stevens Publishing, [2021] | Series: Engineering our world | Includes index.
Identifiers: LCCN 2019030868 | ISBN 9781538247112 (paperback) | ISBN 9781538247129 | ISBN 9781538247136 (library binding) | ISBN 9781538247143 (ebook)
Subjects: LCSH: Artificial satellites–Juvenile literature.
Classification: LCC TL796.3 .B37 2020 | DDC 629.43–dc23
LC record available at https://lccn.loc.gov/2019030868

First Edition

Published in 2021 by
**Gareth Stevens Publishing**
111 East 14th Street, Suite 349
New York, NY 10003



Designer: Sarah Liddell
Editor: Monika Davies

Photo credits: Cover, pp. 1, 5 Andrey Armyagov/Shutterstock.com; background Jason Winter/Shutterstock.com; p. 7Torquemada082/Wikimedia Commons; p. 9 FREDERIC J. BROWN/Staff/AFP/Getty Images; p. 11 BotMultichillT/Wikimedia Commons; p. 13 Science & Society Picture Library/Contributor/SSPL/Getty Images; p. 15 China News Service/Contributor/Visual China Group/Getty Images; p. 17 aapsky/Shutterstock.com; p. 19 image courtesy of European Space Agency; p. 20 (sponge) Nattika/Shutterstock.com; p. 20 (scissors) IB Photography/Shutterstock.com; p. 20 (duct tape) Bjoern Wylezich/Shutterstock.com; p. 20 (construction paper) MNI/Shutterstock.com; pp. 20, 21 (metal nut) Davydenko Yuliia/Shutterstock.com; p. 21 (toothpick) doomu/Shutterstock.com; p. 21 (cup) Sherra/Shutterstock.com.



Printed in the United States of America



CPSIA compliance information: Batch #CS20GS: For further information contact Gareth Stevens, New York, New York at 1-800-542-2595.

# Contents



Words in the glossary appear in **bold** type the first time they are used in the text.

# Machines in the Sky!

High above our heads, machines called satellites **orbit** Earth. These machines collect **data** and then send that data back to Earth. The data we get from satellites helps improve our lives. For example, if you've watched a weather report on the news, you've seen how data from satellites is used to **predict** the weather!

It isn't easy getting a machine up into Earth's orbit. Engineers work hard to plan, build, test, and send satellites into space. It isn't cheap either! Sending a satellite to space costs millions of dollars.

## Building Blocks

Engineers are people who use science and math to build better objects. Engineers who work on spacecraft, like satellites, are called aerospace engineers.

SOME SATELLITES COLLECT AND SEND DATA USED FOR THE GLOBAL POSITIONING SYSTEM (GPS), WHICH LOCATES PLACES ON EARTH. THIS IS THE **TECHNOLOGY** IN OUR MAP APPS THAT HELPS US FIND WHERE WE ARE!

# Making a Plan

Years before a satellite is sent into space, engineers plan out how it will work. They consider what technology the satellite will need, because once it's in space, they can't make changes. For example, if a satellite is needed to predict the weather, engineers add cameras and **sensors** to collect data about Earth's **atmosphere**.

Engineers also think about what to use to make the satellite. What the satellite is made of has to handle the sun's powerful heat and be strong enough to withstand a rocket **launch** into space.

## Building Blocks

Engineers look for light, long-lasting materials, or matter from which something is made, to build satellites. The weight of a material is important because the heavier a satellite is, the more expensive it is to send into space.

NASA SOMETIMES USES GOLD AS A THIN COVER TO COAT SATELLITE PARTS. GOLD THROWS BACK LIGHT AND PROTECTS THE SATELLITE AGAINST THE SUN'S RAYS.

# Building Subsystems

Once engineers come up with a plan for a new satellite, they begin building it here on Earth. Satellites have groups of parts that work together. Each group of parts is called a subsystem. For example, a satellite's **communication** subsystem includes all the parts that send and receive data from Earth.

Since each subsystem takes a lot of work to build, a team of engineers may work on only one part of the satellite at a time. This allows the engineers to make sure each subsystem works perfectly.

## Building Blocks

Satellites have a subsystem called the attitude and orbit control system (AOCS). "Attitude" describes the direction a satellite is facing. The AOCS subsystem keeps the satellite in the correct orbit and its antennas pointing toward Earth.

THESE ENGINEERS ARE WORKING ON A SUBSYSTEM FOR A SATELLITE THAT WILL OBSERVE SOIL MOISTURE AROUND THE WORLD. THIS SUBSYSTEM IS BEING BUILT FOR PROPULSION, OR TO DRIVE THE SATELLITE FORWARD.

# Putting It Together

Once all the subsystems of a satellite are built, engineers bring everything to one place and put together the final machine. This process, or series of steps, is called assembly, integration, and testing (AIT).

Since each satellite is usually unique, or one of a kind, engineers must test every subsystem individually. They also make sure each part of the satellite is properly connected to the rest of the machine. It can take around a year before engineers decide that the satellite is ready to be launched into space.

## Building Blocks

Engineers can become experts in AIT. An expert is someone who knows a great deal about something. AIT experts are skilled at putting together satellites and testing all the subsystems to make sure they're working correctly.

TO TEST SATELLITES THAT ARE FULLY PUT TOGETHER, ENGINEERS PLACE THE MACHINES IN HUGE ROOMS THAT RE-CREATE, OR COPY, SOME OF THE CONDITIONS FOUND IN SPACE.

# Testing, Testing: 1, 2, 3!

After the satellite is built, engineers put it through final tests. This might seem like *a lot* of testing, but it's important to know if the satellite is ready. These tests include checking to see if the satellite's **batteries** are working and whether the satellite can send and receive data.

The satellite is then tested to see if it will survive the launch into orbit. One way to check this is a **vibration** test, which shakes the satellite to make sure it can handle a ride on a rocket.

## Building Blocks

Sending a satellite into space is expensive. The launch itself costs between $10 million and $400 million! Due to this high price tag, it's worthwhile for engineers to test every single part completely.

IT'S ALSO IMPORTANT TO TEST HOW SATELLITES WILL HANDLE THE VERY HOT OR COLD CONDITIONS OF THEIR LAUNCH INTO SPACE.

# Ready for Liftoff!

It's finally time to launch the satellite into orbit! Contents that are taken into space by a rocket are called payloads. In this case, the payload is the satellite, which is placed inside the rocket near the top. Once the rocket leaves Earth's atmosphere, the payload safely breaks away from the main rocket, and the satellites are then able to move into orbit.

As a rocket launch is very expensive, rockets often carry more than one satellite. Since 1957, over 8,000 satellites have been launched into orbit.

## Building Blocks

Sputnik I was the first human-made satellite successfully launched into space. A country called the Soviet Union sent up the soccer-ball-sized machine on October 4, 1957.

ROCKETS HAVE BEEN ABLE TO BRING SATELLITES THAT WEIGH AS MUCH AS SEVERAL TONS INTO SPACE.

# Powering Satellites in Orbit

When satellites are in orbit, they need power to transmit, or send, the data they collect. Satellites are often powered by solar panels, or broad, flat instruments that **convert** the energy given off from the sun into electrical energy. This energy is then stored in rechargeable batteries, which power satellite systems.

Solar panels are also found in everyday tools, such as thermometers, or instruments that measure how hot or cold something is. The solar panels on satellites are much larger than ones in a thermometer, but they work in a similar way.

## Building Blocks

On March 17, 1958, the Vanguard 1 satellite launched into space and became the first satellite to use solar panels as a power source. It collected and sent data until 1964.

EACH BLUE RECTANGLE ON THIS SATELLITE IS A SOLAR PANEL. SOLAR PANELS FACE TOWARD THE SUN SO THAT MORE LIGHT CAN HIT THE PANELS TO CREATE ELECTRICITY.

# Shutting Down Old Satellites

A satellite can only last so long. Over time, old satellites run out of the fuel, or power, needed for certain subsystems.

Engineers have two options before satellites run out of fuel. The first is to send the satellite farther into space to make room for new satellites. The second is to send the satellite back toward Earth's surface. On its way, the satellite usually burns up in Earth's atmosphere. If the satellite's too big to burn up, engineers guide it to fall in the ocean, away from people.

## Building Blocks

Sending satellites farther out into space adds to the space junk already orbiting Earth. This **debris** may ruin new satellites sent into space. Engineers are working to clean up the space junk, but it isn't an easy problem to fix.

THIS IMAGE FROM NASA SHOWS HOW MUCH SPACE JUNK IS ORBITING EARTH. MOST OF THE WHITE DOTS ARE OLD SATELLITES THAT DON'T WORK ANYMORE.

# Make Your Own Satellite

Satellites may be far above our heads, but we can also make a satellite model right here on Earth!

**What You Need:**

- juice box or clean sponge
- scissors
- duct tape or any shiny tape
- glue
- nuts, bolts, or paper clips
- pipe cleaners or popsicle sticks
- toothpicks or pipe cleaners
- construction paper
- small throwaway bowl or cup
- thimble or a metal nut

## How To Do It:

1. Start with the main body of your satellite. Wrap the juice box or sponge in construction paper and tape it all together.
2. Create your power source. Cut out pieces of cardboard or paper and cover them in shiny tape to look like solar panels. Use tape or toothpicks to attach them to the main body.
3. Use tape or glue to attach various "instruments," such as screws or paper clips, to the main body. These instruments collect data for your satellite.
4. Build a communication device using a throwaway bowl or cup. This device is often shaped like a rod or dish. Add the device to the main body using toothpicks, tape, or glue.
5. Choose the satellite's orientation finder, such as a thimble or metal nut. This device lets your satellite know if it's facing the right direction. Then, attach it to the main body. Your satellite is complete and ready for space duty!

# Glossary

**atmosphere:** the combination of gasses that surround a planet, such as Earth

**battery:** a device that turns chemical energy into electricity

**communication:** the sending of a message containing data

**convert:** to cause to change form

**data:** facts and figures

**debris:** the remains of something that has been broken

**launch:** to send out with great force. Also, the act of sending something out with great force.

**orbit:** to travel in a circle or oval around something, or the path used to make that trip

**predict:** to guess what will happen in the future based on facts or knowledge

**sensor:** a tool that can spot changes in its surroundings

**technology:** the way people do something using tools and the tools that they use

**vibration:** a speedy movement back and forth

# For More Information

## Books


Fretland VanVoorst, Jenny. *Artificial Satellites.* Minneapolis, MN: Pogo, 2017.

Graham, Ian. *You Wouldn't Want to Live Without Satellites!* New York, NY: Franklin Watts, 2019.

Woolf, Alex. *The Science of Spacecraft: The Cosmic Truth about Rockets, Satellites, and Probes.* New York, NY: Franklin Watts, 2019.


## Websites

**NASA Knows! for Students K-4**
*www.nasa.gov/audience/forstudents/k-4/stories/nasa-knows/index.html*
Read all about different space topics, including information about satellites.

**Science Podcasts for Kids: The Launch of the Irish Satellite**
*www.sciencepodcastforkids.com/single-post/2019/04/19/The-Launch-of-the-Irish-Satellite*
Listen to a podcast about the launch of Ireland's first satellite.

**What Can Satellites Do?**
*www.wonderopolis.org/wonder/what-can-satellites-do*
Discover more about how satellites improve our lives here.

**Publisher's note to educators and parents:** Our editors have carefully reviewed these websites to ensure that they are suitable for students. Many websites change frequently, however, and we cannot guarantee that a site's future contents will continue to meet our high standards of quality and educational value. Be advised that students should be closely supervised whenever they access the internet.

# Index